MW00763461

FEMALE FIRSTS IN THEIR FIELDS

FEMALE FIRSTS IN THEIR FIELDS

AIR & SPACE

Doug Buchanan

Introduction by
Roslyn Rosen

CHELSEA HOUSE PUBLISHERS
Philadelphia

Produced by P. M. Gordon Associates, Inc.
Philadelphia, Pennsylvania

Editor in Chief Stephen Reginald
Managing Editor James D. Gallagher
Production Manager Pamela Loos
Art Director Sara Davis
Director of Photography Judy L. Hasday
Senior Production Editor Lisa Chippendale
Publishing Coordinator James McAvoy

Picture research by Gillian Speeth, Picture This
Cover illustration by Cliff Spohn
Cover design by Keith Trego

Frontispiece: Christa McAuliffe

The Chelsea House World Wide Web site address is
http://www.chelseahouse.com

First Printing

1 3 5 7 9 8 6 4 2

Library of Congress Cataloging-in-Publication Data

Buchanan, Douglas.
    Female firsts in their fields. Air and space / Douglas Buchanan, Jr.
      p. cm.
    Includes bibliographical references and index.
    Summary: Profiles women who have been active in air and space,
including Christa McAuliffe, Amelia Earhart, and Sally Ride.
    ISBN 0-7910-5141-2 (hardcover)
    1. Women air pilots–Biography–Juvenile literature. 2. Women
Astronauts–Biography–Juvenile literature. [1. Air pilots.
2. Astronauts. 3. Women–Biography.] I. Title.
TL539.B83 1998
629.1′2736–dc21
   [B]   98-45723
                                            CIP
                                            AC

CONTENTS

INTRODUCTION

Roslyn Rosen

When I was a toddler, it struck me that the other people in my family's New York apartment building were different. They did not use their hands when they talked, and they did not have to watch each other speak. I had been born deaf, and I felt sorry for them because they did not know the joy of drawing pictures in the air. They could not splash ideas into the air with a jab of the finger or a wave of the hand. Not until later did I realize the downside of being deaf–I couldn't communicate directly with my grandparents and extended family members, I depended on others to make important phone calls for me, and I found life's opportunities narrower, in part because I had few deaf (let alone female) role models.

Gallaudet University in Washington, D.C., is the only college for deaf students in the world. I arrived there in September 1958. It was a haven where sign language was part of the educational process, where there were deaf professors, and where opportunities for extracurricular leadership abounded. At Gallaudet I met deaf female professionals for the first time, although there were probably not more than three or four. The president and administrators of Gallaudet were all males who could hear–typical of school administrations during those years.

In my first month at Gallaudet, I also met the man who would become my husband. My destiny was charted: major in something that I could use as a homemaker (since that would be my job), get married, have a bunch of kids, and live happily ever after. This was the expectation for women in the late 1950s and early 1960s. And I stuck to the script: I majored in art with an emphasis on education and English, got married, and had three children. My life was complete–or so I thought.

The 1960s were turbulent and thought-provoking years. The civil rights movement and the beginnings of a women's movement emphasized human rights and equality for all. I came to see how alike the issues were that faced women, people of color, and people with disabilities, in terms of human rights and respect for human differences. Multicultural studies are vital for this understanding. Changes were occurring at an accelerating rate. Those changes affected my husband and me by broadening our traditional gender roles. With my husband's support, I pursued a master's degree in education of deaf students and later a doctoral degree in education administration. From my first job as a part-time sign language teacher, I eventually joined the faculty at Gallaudet University. In 1981 I was promoted to dean of the College for Continuing Education, and in 1993, to vice president for academic affairs.

During the formative years of my career, many of my role models and mentors were deaf men who had reached positions of leadership. They hired, taught, advised, and encouraged me. There were times when I felt the effects of the "glass ceiling" (an invisible barrier that keeps women or minorities from rising any higher). Sometimes I needed to depend on my male colleagues because my access to "old boy" networks or decision makers was limited. When I became involved with the National Association of the Deaf (NAD), the world's oldest organization of deaf people, I met deaf women who became role models–Dr. Gertie Galloway was the first deaf female president of the NAD, and Marcella Meyer had founded the Greater Los Angeles Community Service of the Deaf (GLAD). In 1980 I was elected to the board of directors of the National Association of the Deaf, and in 1990, I became the second woman elected president of NAD.

When I became a dean at Gallaudet in 1981, I also became a member of the school's Council of Deans, which at the time included only two deaf deans and two female deans. I was the only deaf woman dean. The vice president was a white male, and he once commented that top administrators often build management teams in their own image. I have found that to be true. As a dean, I was the highest-ranking deaf woman at Gallaudet, and I was able to hire and help a number of young deaf female professionals within the College for Continuing Education and our regional centers around the country. In the five years that I have been vice president at Gallaudet I have added many deaf, female, and minority members to my own management team. When I was the president of NAD, I hired its first deaf female executive director, Nancy Bloch. I also encouraged two of my friends, Mabs Holcomb and Sharon Wood, to write the first deaf women history book, a source of inspiration for young deaf girls.

It is important for women who have reached the top levels of their fields to advise and help younger women to become successful. It is also important for young girls to know about the groundbreaking contributions of women who came before them. The women profiled in this series of biographies overcame many obstacles to succeed. Some had physical handicaps, others fought generations of discriminatory attitudes toward women in the workplace. The world may never provide equal opportunities for every human being, but we can all work together to improve life for the next generation.

DR. ROSLYN ROSEN is the Vice President for Academic Affairs at Gallaudet University in Washington, D.C. Dr. Rosen has served as a board member and President of the National Association of the Deaf (NAD), the oldest consumer organization in the world, and was a member of the National Captioning Institute's executive board for nine years. She is currently a board member of the World Federation of the Deaf. Dr. Rosen also wears the hats of daughter, wife, mother, and proud grandmother.

AMELIA EARHART

Amelia Earhart strained to see the Atlantic Ocean far below her, but thick fog obscured the cold, wave-tossed sea. Her window, tiny and frosted, offered only a view of the Fokker airplane's long sturdy wing, and even that was sometimes hidden by menacing gray clouds sliding past. Snow swirled around the airplane, churned by the Fokker's three rumbling engines. Occasionally hail battered the airplane's plywood skin. From inside, it sounded as though the Fokker was flying through a downpour of gravel.

Bundled up against the intense cold, her head pounding from the roar and vibration of the engines, Amelia Earhart aimed a borrowed camera out her small window, snapping pictures of passing cloud formations. Despite the cold, the noise, and the angry weather, there was no place Amelia Earhart would have rather been on June 18, 1928, than in this airplane far out over the wide Atlantic. At 30, she had been a pilot for six years, but she was not at the controls of the big Fokker, nicknamed the *Friendship*, as it sped across the ocean toward Europe. For this trip, she was only a pas-

Amelia Earhart in the cockpit of an autogiro in 1931, after setting a women's altitude record for this type of plane.

senger, but Earhart was making history just the same. When the Fokker reached England, Amelia Earhart would become the first woman ever to fly across the Atlantic Ocean.

Finding England in such dense fog was not easy. In the Fokker's cramped cockpit, pilots Wilmer Stultz and Louis Gordon sat shoulder to shoulder, intent on bringing the plane safely to Southampton, their final destination. Flying across the ocean in 1928 was very dangerous; dozens of skillful pilots and their frail planes had disappeared into the freezing North Atlantic, never to be heard from again. Many had died trying to duplicate the heroic feat of Charles Lindbergh, who had flown his single-engine plane from New York City to Paris without stopping.

Serious aviators did not consider these flights to be stunts, or adventures for adventure's sake. Such flights were meant to prove that airplanes were capable of flying long distances, even through terrible weather. In the future, it was hoped, airplanes would be able to safely carry passengers and cargo anywhere in the world. At the time of Amelia's first transatlantic flight, however, this dream was far from reality. In 1928 alone, 14 aviators had perished trying to guide their planes over the ocean to Europe from America. By joining the crew of the *Friendship*, Amelia Earhart was risking her life.

Several times during the long flight, Earhart traded places with Stultz or Gordon to spend time in the cockpit, the only heated part of the big Fokker. As she peered into the icy mists surrounding her, Earhart shrugged off the effects of numbing cold and vibration, but she could not entirely shake off the frustration caused by the foul weather. "We might as well have been flying over the cornfields of Kansas," she said later, for she had been troubled

In July 1928, Chicago admirers welcome Earhart, who has just become the first woman to fly across the Atlantic. Framing her are pilots Wilmer Stultz (left) and Louis Gordon.

that the ocean—and even England itself—lay hidden far below.

Perhaps Earhart also felt frustration of a different kind—the frustration any pilot would feel making a historic flight as a mere passenger. Although she had been permitted to sit in the cockpit, she had not been given the chance to fly the Fokker herself. No doubt Earhart wanted to take command of the plane and use her own skills as a pilot to guide the *Friendship* to England. But it was not to be, at least not on this trip. Yet Amelia Earhart would fly across the Atlantic Ocean again, and she would do it alone.

The Fokker flew on through fog, ice, and snow for nearly an entire day, until at last, Earhart, Stultz, and Gordon saw land below them. They realized the *Friendship* would run out of gas before they could reach Southampton. Circling a small seaside town, Stultz landed the *Friendship* at Burry Port, Wales. With the ocean behind them and the flight a success, Amelia faced a new challenge: seeing her dreams come true.

Amelia Earhart had not always wanted to be a pilot. When she saw her first airplane at age 11, the frail machine had not impressed her. But even as a young girl, Earhart suspected she had a special path in life. She was born in Atchison, Kansas, on July 24, 1897, at the home of her grandparents, Alfred and Amelia Otis. Her father, Edwin Earhart, worked as a lawyer for a nearby railroad throughout much of Earhart's childhood. Edwin Earhart was at times a gentle and caring father. Yet he battled a drinking problem for many years, sometimes making life in the Earhart family very difficult.

Amelia's mother, Amy, tried to give her two daughters a sense of independence and confidence. With her younger sister Muriel, Amelia was always in search of adventure, but sometimes she found trouble instead. Many of the pastimes she enjoyed—fishing, hiking, and football—were considered boys' games. Amelia participated in them anyway.

By her 17th birthday, much had changed for Amelia since her childhood. Her father's drinking had forced her parents to separate, and Amelia moved with her mother and sister to Chicago, where she entered high school. Learning interested her a great deal, but grades did not. After graduation, she attended Ogontz, a small college for women in Philadelphia.

Few of Earhart's classmates shared her desire for adventure and accomplishment. Many simply wished to finish college, find husbands, and raise families. At that time, few women wanted professional careers in the way Amelia did. Yet more than ever, she was determined to follow her own path in life.

In fall 1919, Amelia Earhart began to study medicine at Columbia University in New York City. There she worked harder than nearly any of her classmates. During her year at Columbia, Amelia's parents had reunited and moved to Los Angeles. In summer 1920, they asked Amelia to give up her studies at Columbia and to live with them in California. The decision to leave college was a difficult one, but it did take her to California, and there she would again meet the machine that had left her so unimpressed at age 11, the airplane.

The many young U.S. pilots returning home from Europe after the First World War often found little to do with their flying skills. Many earned a living by performing dangerous and exciting stunts at air shows across the country. Southern California, with its warm days and clear skies, was an especially popular location for air shows. After attending several of the spectacular shows while living with her parents in Los Angeles, Amelia Earhart fell in love with flying.

Amelia Earhart in the early 1920s when she applied for a pilot's license.

Her first ride in an airplane lasted only 10 minutes, but Amelia was hooked. "I knew I had to fly," she later wrote. One night at the dinner table, Amelia announced to her family she intended to become a pilot. Her parents were used to Amelia's adventurous ideas. But flying was an expensive hobby, too expensive for the Earhart family. Amelia refused to let the family's poverty stand in the way of her new dream. She worked long hours at odd jobs to pay for her flying lessons. After two years of saving and flying, Amelia Earhart was awarded her pilot's license. She was ready to enter the exciting–and dangerous–world of aviation.

As a present to herself for her 24th birthday, she

bought her first airplane, a used Kinner Airster, for the then-tremendous sum of $2,000. The little Airster was painted bright yellow, so Earhart nicknamed it "the Canary." Although she had become an able pilot, the Airster could be a difficult plane to fly. Its motor had a bad habit of quitting in the air.

Still, Earhart set her very first world record in the Airster, pushing the airplane to the incredible altitude of 14,000 feet, higher than anyone had ever flown before. She might have flown higher still had her motor not begun coughing and wheezing, as if to protest being so far off the ground. But Earhart understood the dangers of flying very well. Landings could be especially hazardous. She had once flipped the Airster completely over on its back while trying to land, the force of the crash throwing her from the open cockpit onto the ground.

While Amelia continued to gain flying experience, her family's money situation worsened. A series of poor investments had left her parents without income, and the strain of poverty finally finished off Amy and Edwin's troubled marriage. Amelia's mother, wishing to return to the East Coast with her daughters, persuaded Amelia to sell the Airster and to drive her back east in a newly purchased car.

Leaving California and the Airster behind, Amelia went back to her studies at Columbia, but money troubles soon forced her to take a job as a social worker in Boston. Though social work did not offer the excitement of stunt flying, Amelia enjoyed helping people in need. She had battled poverty herself and knew the value of a helping hand during difficult times. Still, she flew whenever she could. She met many of Boston's best-known pilots by joining an important flying organization, the National Aeronautic Association.

Soon Amelia Earhart's name was as well-known

among pilots in Boston as it had been in California. Her hard work paid off when she was selected to join the crew of the *Friendship* for their flight across the Atlantic Ocean. When she came home from her adventure with Bill Stultz and Louis Gordon, Earhart expected to return to the life of a social worker. But to many Americans, she had become a hero. Many thought Amelia had proven that women could withstand the hardships of long-distance flying. Of course, Earhart had known this all along.

Amelia had never expected the welcome she received after returning home from England. Cheering crowds followed her nearly everywhere she went. Few people were interested in the men; the public wanted to see Amelia. Amelia's picture and her exciting story were featured in newspapers around the world. Her adventurous spirit, her courage, and her beauty were admired by millions. Amelia decided she could not return to social work: she would devote her life to flying.

With fame came money, and Amelia used both to her advantage. On July 30, 1929, she purchased a new airplane, a Lockheed Vega. The big Vega was everything the Canary was not; it was sturdy and fast, with a powerful engine. Almost immediately, Amelia set out to test both the Vega and her flying abilities. She had already piloted another plane, an Avro Avian, to become the first woman to fly across the United States and back, but the Vega opened up new possibilities. With its impressive speed and ability to fly long distances, Amelia's Vega was an excellent racing machine. Competing against the finest female pilots in America, Amelia raced her Vega from California to Ohio in the 1929 Women's Air Derby, finishing in third place. With her new experience in long-distance flying, Amelia set out

Amy Otis Earhart, Amelia's mother, traces her daughter's progress on an atlas as she listens to radio reports in March 1937.

to tackle her greatest challenge yet. She would fly across the Atlantic Ocean again—this time on her own.

With the Vega loaded with gasoline, Earhart took off for Europe from Newfoundland, Canada, on May 20, 1932. For several hours, she guided the big Vega through the clouds without trouble. Then things began to go wrong. Her altimeter quit, so she could no longer tell exactly how high she was above the wind-whipped ocean. Soon the Vega's big engine began to falter. The weather worsened, the air grew cold, and a heavy coat of ice formed on the plane.

Earhart worried she might not make it. She flew on through the night without resting.

By dawn, Earhart was relieved to see green pastures, farms, and towns below her. Spotting a grassy field, she landed safely and climbed out. An astonished farmer ran toward her. She called out, asking where she had landed. The farmer's thick Irish accent answered her question. Amelia Earhart had conquered the Atlantic alone.

With the Vega repaired, Earhart made other record-breaking flights. She became the first pilot to fly solo from Hawaii over the Pacific Ocean to California, then the first to fly alone to Mexico City from Los Angeles. Earhart flew across the United States again in just 17 hours and 7 minutes, breaking her own previous record. She seemed to be able to accomplish anything, and her fame was greater than ever before. People everywhere cheered Amelia on. By 1937, Earhart decided the time was right for the greatest challenge of her career as a pilot: she would fly completely around the world.

To accomplish this, Earhart was given a special plane. Her shiny new Lockheed Electra was bigger than the Vega and had two engines. If one quit, she could land by using the other. There was always the danger of running out of gas far from any airfields or towns, but Earhart planned her trip with care. Unless she became lost, she would have enough fuel for each part of the trip.

Amelia's husband, George Palmer Putnam, helped her plan her route. They spent many days together studying Amelia's collection of maps and decided Amelia should start in Florida and fly east across the Atlantic to Africa. She worried most about flying over Africa. The thick jungles offered few places to land if anything went wrong. Leaving Africa, she would fly to India, then to Australia, then across the

Earhart and Fred Noonan board their Lockheed Electra in San Juan, Puerto Rico, an early stop on their fateful trip.

Pacific Ocean back to America. Many of Earhart's friends worried most about her flying across the Pacific, a much wider ocean than the Atlantic. But Earhart would not be alone on this trip. She was glad to have Fred Noonan, an experienced navigator, to help her find her way around the world.

Heavy with extra fuel, the Electra took off from Miami, Florida, on June 1, 1937. For more than a month, Earhart and Noonan flew from country to country, lofting over oceans, mountain ranges, deserts, towns, and cities. At every stop, tremendous crowds welcomed them. As they reached Australia, Earhart began to worry about flying across the wide Pacific. The Electra would need to stop to refuel twice, once at tiny Howland Island, and once in Hawaii. As Earhart knew, finding Howland Island in the middle of the vast ocean would require great skill. Anchored off Howland was an American ship, the *Itasca*. Earhart and Noonan would follow radio signals from the *Itasca* to locate the island.

The world held its breath as the Electra took off at 10 in the morning from New Guinea to begin the first leg of the Pacific crossing. Early the next morning, radiomen aboard the *Itasca* heard Earhart approaching Howland Island. But somewhere over the dark ocean, after many hours in the air, the Electra had wandered off course. Flying through rain clouds, Earhart and Noonan could not find Howland or the *Itasca*. The sailors aboard the *Itasca* became extremely worried. They tried again and again to contact the Electra over the radio, desperately trying to guide Earhart to Howland Island. But they could do nothing more to help her. By noon, they knew it was too late. The Electra must have run out of fuel and crashed into the ocean. Amelia Earhart and Fred Noonan were gone.

Although many searches were made, no trace of Earhart, Noonan, or the Electra was ever found. People around the world were stunned to hear that Earhart's greatest flight had ended in disaster. Yet her achievements as a pilot and as a woman have endured. She is honored for her contributions to aviation, and for her courage and daring in the cockpit. Perhaps more important, Amelia Earhart is revered for her role in the struggle for equality between men and women. Many Americans protested that women had no place in the skies, and that flying was a man's business. With an airplane throttle at her fingertips, and a wide open sky ahead, Amelia Earhart, pioneer aviator, knew better.

SALLY KRISTEN RIDE

With a heavy *thwop!* Sally Ride pounded the tennis ball back over the net. Her opponent lunged for the ball but missed, falling back in defeat. At 11 years old, Sally Ride thought she might someday become a professional tennis player. So did the tennis stars who came to watch her play. Several even encouraged her to leave school and devote herself full time to tennis. By her teenage years, Ride was ranked as the 18th best junior tennis player in the nation. She excelled at nearly every sport she tried, including baseball, football, and rugby. She loved the thrill of competition. In sports, winning seemed to come naturally.

Sally Ride was born on May 26, 1951, in Encino, California. She grew up with her younger sister Karen, nicknamed "Bear." Sally and Bear thrived in an encouraging household. Their father, Dale, a professor of political science at a nearby community college, taught them to excel at everything they tried. Sally's mother, Joyce, stayed at home, where she kept her two adventuresome girls out of trouble.

Sally Ride, soon to become the first American woman in space, arrives at Kennedy Space Center in Cape Canaveral, Florida, in June 1983.

Preparing for her historic mission, Ride visits the flight line at Ellington Air Force Base, Texas, in 1982.

Sally was as talented in the classroom as she was on the tennis court or the rugby field. She studied with great determination. Her efforts earned her a partial scholarship to the prestigious Westlake School, an all-girls high school in Los Angeles. At Westlake, Ride found a special friend and mentor in Elizabeth Mommaerts, her physiology teacher. There were other science subjects that interested Sally more than the study of bodily processes, but by watching Mommaerts, Sally Ride began to understand how scientists confronted difficult problems. Ride's lessons in "the scientific method" taught her to use logical steps to answer challenging questions about the world around her.

At Westlake, Ride continued to develop her love for science through classes in chemistry, physics, and mathematics. In 1968, she went to Swarthmore College outside Philadelphia, where she continued her studies in physics but found she greatly missed the thrill of athletic competition. After only three semesters, Ride left Swarthmore to devote her life full time to tennis. But as great as her natural talents were, she finally decided she did not have the skills to become a tennis pro. By 1970, Ride was back in school.

When she arrived at Stanford University in California in 1970, one of the finest universities in the United States, Ride tackled her studies with her typical determination. Although most students majored in one subject, Ride took on the extra work required to major in two. When she had completed her undergraduate studies, she received degrees in both physics and English literature. Although Ride had always been fascinated by science, she had come to love the plays of William Shakespeare. Her literature classes gave her a chance to study Shakespeare and to take a break from science.

Her grades continued to be excellent. After finishing her bachelor's degrees, Ride set out to earn graduate degrees. She considered studying Shakespeare but decided to stick to her first love, science. As a master's student and doctoral candidate at Stanford, Ride focused on the field of astrophysics, the study of the movement and behavior of objects in space. This study would prove particularly important to Ride's later career as an astronaut.

While she finished her last papers and took her final exams as a doctoral student, an advertisement in the Stanford University newspaper caught Sally's eye. The National Aeronautics and Space Administration, known as NASA, was looking for future astronauts. With her background in space science, and her long list of personal achievements, Sally thought she had a good chance of being selected.

Sally's application was only one of more than 8,000 received by NASA in the following months. Her outstanding work as a graduate student won her a position among 208 finalists—only 35 would be selected to become future astronauts. NASA's selection process was intense. Only the best men and women in their fields would be chosen for the long and expensive astronaut training program. Ride was interviewed by psychiatrists, poked and prodded by doctors, and questioned by agents from the FBI. But she remained optimistic. She knew her hard work had paid off when she received a call from NASA informing her that she'd been chosen. Sally Ride, now Dr. Sally Ride, would become an astronaut. Four other women were also chosen—a surgeon, a biochemist, a geologist, and an electrical engineer.

The long interview process was only the beginning. To become an astronaut, Ride had to get to know her new spacecraft inside and out. NASA's

On the hot seat: during training at a "survival school" in 1978, Ride positions herself on a device that simulates ejecting a pilot from an aircraft.

newest spaceship in 1978 was the space shuttle, which had not yet flown in space. While NASA continued to test the space shuttle on Earth, Sally set off on the long road to becoming an astronaut. The training required every bit of the discipline Sally had built up as a star tennis player and physics student. She rode in a special NASA plane to experience weightlessness, practiced her new piloting skills, and studied navigation, space propulsion, and aerospace engineering, among many other subjects. She and her fellow astronaut trainees spent hours in the space shuttle simulator, memorizing the locations and functions of the thousands of switches, knobs, and dials that control the shuttle.

Sally was also assigned to a very important project. The space shuttle uses a long robotic arm to move satellites into and out of its cargo bay. Known to astronauts as the remote manipulator system, or RMS, the arm is a crucial part of the space shuttle. Without it, astronauts working outside the shuttle would have a very difficult time moving large objects. As part of a special team, Sally was assigned the task of designing the RMS. She would later use the arm in space herself.

By November 1981, the space shuttle was ready to fly its first mission. Although Sally Ride was not among the crew of the first two flights, she was assigned the important role of capsule communicator, or CAPCOM. As CAPCOM, Ride radioed messages, instructions, and questions from her desk at Mission Control in Houston to the shuttle orbiting above. Only the CAPCOM could speak directly to the astronauts during a mission.

Sally Ride's first chance to put her years of careful training to use in space came in June 1983. More than a year earlier, NASA had assigned Ride to fly on the seventh space shuttle mission. Since then, she had been practicing for her mission nearly every day. Long before dawn on June 18, 1983, Sally found herself climbing out of bed in the astronaut's crew quarters at Cape Canaveral, Florida. With her fellow crew members Robert Crippen, Frederick Hauck, John Fabian, and Norman Thagard, Sally Ride boarded the space shuttle *Challenger*. Technicians and engineers had been up all night putting the last touches on preparations for the shuttle's week-long journey. Behind the crew compartment were two multimillion-dollar communications satellites. It would be the crew's job to deliver the satellites safely to space.

At seven o'clock in the morning, *Challenger* blasted off from Cape Canaveral's launchpad 39A, bound for space. Safely strapped into her seat, Sally Ride found herself rocketing into orbit at supersonic speed. *Challenger* performed flawlessly. Once in space, Sally had little time to reflect on her new status as America's first woman in space. Besides the lengthy preparations needed to launch the two satellites, Sally had some 40 scientific experiments to carry out.

The crew's first task was to deploy one of the two communications satellites. After opening *Challenger*'s massive cargo bay doors, Sally used the robotic arm to nudge *Anik-C,* a Canadian communications satellite, into orbit. America's first woman in space later launched the Indonesian satellite *Palapa B-1,* the second of the two communications

The Challenger *crew wave as they leave for boarding on June 18, 1983. Ride and Robert Crippen lead the way; behind them is John Fabian, with Norman Thagard and Frederick Hauck in the rear.*

satellites. *Palapa B-1* would help provide telephone service to millions of Indonesians. The crew's scientific experiments investigated, among many other topics, the possible causes of space sickness, a nausea that sometimes bothers astronauts during their first few days in weightlessness.

After six days and 98 orbits around the Earth, *Challenger* made a perfect landing at Edwards Air Force Base in California. Commenting on the success of her first flight into space, Ride told admiring reporters that she'd had "the most fun I will ever have in my life." After returning to Earth, Sally enjoyed answering the many questions about her experience. She was, however, quick to correct people who referred to her as the first woman in space, since that honor belonged to the Russian cosmonaut Valentina Tereshkova.

Sally Ride would have another chance to put her training as an astronaut to the test. On October 5,

1984, Sally boarded *Challenger* yet again, this time for her second mission to space. *Challenger* was more crowded than ever before, carrying a full load of seven astronauts. Sally was not the only woman aboard *Challenger* this time. With her was astronaut Kathryn Sullivan, who would make the first spacewalk by an American woman. As on her first mission, Ride helped launch an important satellite and carried out a number of scientific experiments.

Sally Ride's years studying science were paying off. She often thought about Elizabeth Mommaerts, the teacher who had so inspired her years before. In turn, many American women looked to Ride as their role model. She had impressed her colleagues at NASA as well. One of her biggest fans was NASA's director of flight crew operations, George Abbey. "Sally is smart in a very special way," he said.

The *Challenger* disaster in 1986 hit Sally Ride very hard. She watched as the shuttle that had twice carried her into space was destroyed in a tremendous explosion, killing its entire crew. Sally was appointed to a special board of investigation, which was given the task of finding out how the accident had happened.

In 1987, after finishing her investigation into the *Challenger* disaster, and after serving on NASA's long-term planning committee, Sally Ride announced she was retiring from the astronaut corps. She decided to return to where she had started ten years earlier, accepting the prestigious position of science fellow at Stanford University. Later, Ride moved to San Diego, where she became a physics professor at the University of California. Sally Ride continues to encourage space exploration, and to inspire future generations of scientists and astronauts.

SHANNON W. LUCID

Every day, Shannon Lucid paused by the window on the way to her office. While many of her friends at NASA enjoyed office-window views of the Johnson Space Center's neatly trimmed lawns, Lucid's view was a bit more spectacular. Far below her, a brilliant Earth slowly rotated in the sunshine. She could see forest fires burning in Mongolia, clouds drifting across Florida, or watch evening shadows lengthen across the Himalaya mountain chain. After four successful space shuttle missions, Shannon Lucid found herself living in Earth orbit aboard the Russian space station *Mir*.

Unlike her space shuttle missions, which had not lasted longer than two weeks, her mission of scientific research aboard *Mir* was to last more than six months. She would return home having spent more time in space than any other American astronaut, male or female, and more time in space than any woman in the world.

Like several of her fellow astronauts, Shannon Lucid had dreamed of rockets and spaceflight since an early age. She was born in China to missionary parents in 1943. Just six

A day after returning from six months in orbit in 1996, Shannon Lucid waves to earth-bound photographers.

weeks after her birth, Shannon and her parents were captured by the Japanese army, which occupied parts of China during World War II. A prisoner exchange program soon permitted young Shannon to return home to the United States with her parents. She spent most of her childhood in Bethany, Oklahoma, where she developed an acute interest in science. By eighth grade, Shannon was already writing papers on rocket travel.

After her graduation from Bethany High School in 1960, Lucid enrolled at Wheaton College in Illinois, where she began to study chemistry. As a chemistry student, Lucid started to examine the chemical components, or building blocks, of which all things are made. Later, as a graduate student, Shannon focused her studies on the field of biochemistry, which seeks to understand the chemical processes that keep plants and animals alive.

Life as a student was difficult for Shannon. Although she earned top grades, she often found money in short supply. When Wheaton increased its tuition during Lucid's second year, she was forced to return home. She enrolled at the University of Oklahoma, where tuition was more affordable, and finished her bachelor's degree in 1963. During her college years, Lucid fulfilled a longtime dream: she became an airplane pilot. Though her spare time and money were scarce, she took flying lessons and earned her pilot's license. She enjoyed flying so much that she considered becoming a professional pilot for a commercial airline. Yet in the 1960s, most airlines would not hire female pilots. Briefly disheartened, Lucid focused on her love of science, completing her master's and doctorate degrees at the University of Oklahoma.

By the mid-1970s, Shannon had married and was the mother of three children, Kawai, Shandra, and

Michael. Although she found her roles as wife, mother, and scientific researcher rewarding, Shannon Lucid yearned for something more. When NASA began to recruit a new class of astronauts in 1978, Lucid applied. To her astonishment, she was accepted as one of America's first group of female astronauts. By August 1979, Shannon had completed her initial training as a mission specialist.

As a mission specialist aboard the space shuttle, Shannon would carry out complex scientific experiments. During her first spaceflight aboard shuttle *Discovery* in June 1985, she helped perform six important experiments designed by scientists on Earth. Lucid also assisted in several biomedical experiments that examined the human body's responses to the weightless environment of outer space. As a biochemist, Shannon was very interested in seeing if the human body functioned differently in space than on Earth. There were many important questions to be answered. Did weightlessness cause muscles to become weaker over time? Was radiation from the sun more dangerous to people in space than on Earth? Could people live in space for very long periods of time without becoming ill? She would continue to think about important questions like these during her next three space missions.

Between 1989 and 1993, Shannon Lucid rocketed into orbit three more times, twice aboard space shuttle *Atlantis*, and once aboard *Columbia*. Little did Shannon know that her boss at NASA, astronaut Robert Gibson, was considering her for an extremely important space mission. In February 1986, the Soviet Union had begun building what would become the world's largest space station. It was called *Mir*, the Russian word for "peace." Though the Soviet Union had since broken up into

Lucid conducts an inventory of new food supplies with Mir *crewmates Yuri Onufriyenko (left) and Yuri Usachev.*

many smaller countries, the largest, Russia, had continued to use the space station. A special program had been started to allow astronauts from outside Russia to visit the space station. After looking closely at each of his many talented astronauts, Robert Gibson decided Shannon Lucid was ideal for such a mission.

In 1994, Gibson called Shannon to ask if she would like to learn to speak Russian. Shannon immediately knew why. Gibson told her she was being considered for a stay aboard *Mir*, but that the Russians would have the final say. Though she had worked hard as a graduate student and as an astronaut, she knew she would have to work harder than ever to prepare for her mission aboard *Mir*.

Her journey to *Mir* began with a trip to Star City,

the home of the Russian space program. There, Russian cosmonauts trained for their difficult missions in much the same way their American counterparts did in Houston. Shannon attended many demanding classes, all taught in Russian. Her work sometimes allowed her only five hours of sleep each night. Finally, after more than a year's preparation, Shannon Lucid was ready to meet *Mir*.

The world's largest space station, *Mir* took nearly 10 years to build. Since the station was much too large to be carried into space by a single rocket, it was launched into space one piece at a time. In orbit, skilled cosmonauts in space suits had assembled the pieces, called *modules*. Each module was as large as a tractor-trailer truck, and each had its own Russian name, such as Kvant, Kristall, and Spektr. Giant solar fins stretched out from each module, supplying *Mir* with the electricity it needed to run its lights, computers, and life support systems. Since *Mir* would not ever fly through the Earth's atmosphere, it was not streamlined like the space shuttle. Antennae, wires, and equipment poked out in every direction, so *Mir* looked like a "cosmic tumbleweed," Shannon Lucid said.

On March 22, 1996, Lucid boarded the shuttle *Atlantis* for the trip to the "cosmic tumbleweed." Three days later, *Atlantis* safely docked with the giant Russian space station. A special doorway, called an airlock, allowed shuttle astronauts to enter *Mir* without having to wear space suits. When the airlock opened, Shannon was greeted by her new Russian crewmates, Yuri Usachev and Yuri Onufriyenko. The two cosmonauts would run the space station while Shannon conducted her many scientific experiments. While she settled into her new home, *Atlantis* departed, returning with its crew to Earth. According to NASA's schedule, nearly five

Aboard Mir, *Lucid checks on the wheat she grew in a small greenhouse.*

months would pass before another shuttle would arrive to bring Shannon Lucid home.

With so much work to do, the months passed quickly. One of Shannon's most pressing concerns was Mars. Though she would not be traveling to the red planet herself, future astronauts would, and their journey there would take months. Could the fragile human body survive in space for so long? Shannon knew that without gravity, the body's muscles soon weaken. How often would she have to exercise to keep fit in space? Shannon and her fellow cosmonauts jogged on a treadmill every day to make sure their muscles remained strong. Future astronauts might also have to grow their own food in a special type of garden, a garden without soil. Would plants grow normally without gravity? To find out, Shannon grew wheat in a tiny greenhouse.

Though Shannon had many assignments, she also had time to marvel at the incredible views of the Earth below, and to enjoy the company of her fellow space travelers. Yuri Usachev had been in space once before, while Yuri Onufriyenko was making his first space flight. During meals, the three talked about their mission, although often their conversations grew personal. Once, they all admitted that their greatest childhood fear had been that the Soviet Union and the United States would go to war against one another. How silly the idea of war seemed, Shannon thought, considering that aboard *Mir*, Americans and Russians worked side by side in peace.

Though Shannon's mission had been scheduled to last just four and a half months, problems with the space shuttle stretched her mission to more than six months. At last, on September 26, 1996, Shannon Lucid left *Mir* to return to Earth. Including her four space shuttle missions, Shannon Lucid had spent 223 days in space and had set the new world's record for the longest time in space by any woman.

Is Dr. Shannon Lucid finished with outer space? Not just yet. Now back at the Johnson Space Center in Houston, she is preparing for yet another mission. A return to *Mir*, however, is unlikely. According to current plans, *Mir* will soon be replaced by a new and even larger space station, which will be completed over the next 10 years. Given her love for discovery and exploration, Shannon Lucid's suitcase is probably already packed.

CHRISTA McAULIFFE

For a number of U.S. astronauts, their space careers came as complete surprises. When Sally Ride began her physics studies at Stanford, or Shannon Lucid cracked open her first chemistry textbook in Oklahoma, neither imagined they would someday find themselves rocketing into orbit aboard the space shuttle. When Christa McAuliffe began to teach her first class of eighth-grade students at Foulois Junior High School in Morningside, Maryland, she could have hardly imagined that she too would one day join the ranks of America's space professionals.

Sharon Christa McAuliffe was born to Edward and Grace Corrigan on September 2, 1948, in Boston, Massachusetts. With her four younger brothers and sisters, Christa grew up in nearby Framingham, Massachusetts. Grace Corrigan often looked to her eldest daughter to help keep the household running smoothly. At an early age, Christa began passing on her lessons in life to her younger siblings. A teacher was in the making.

After graduating from Marion High School in 1966, McAu-

After a practice session for her space mission, McAuliffe climbs out of a NASA T-38 jet trainer.

liffe studied history at Framingham State College. Four years later, with her bachelor's degree in hand, she was ready to begin her professional teaching career. In 1970 she married Steven McAuliffe, whom she had known since high school, and began teaching at Foulois Junior High School.

Christa made sure her history students understood the important connections between the past, the present, and the future. History was made every day, she told her students. To illustrate this, Christa and her students often talked about history-making current events, such as NASA's moon-landing program. Just the year before, men had walked on the moon for the very first time. Like many of her students, Christa was excited about the exploration of space. She felt the space program could open up exciting future opportunities for her young students.

In the early days of the U.S. space program, only men had been eligible to become astronauts. NASA's first seven astronauts were experienced military pilots, men accustomed to handling dangerous and untested flying machines. By the late 1970s, a new kind of astronaut was being trained for the space shuttle. These new astronauts were more frequently scientists and engineers than military pilots. Their special skills would be needed to conduct the many important scientific experiments to be carried out aboard the new space shuttle.

By the mid-1980s, NASA was ready to train another type of astronaut for a new kind of mission. This new astronaut would not need to be a pilot or a scientist. NASA needed someone with the special ability to talk to young people about space and the future. NASA needed a teacher.

NASA's teacher needed to be intelligent, creative, and physically fit. He or she needed to have shown a lifelong dedication to sharing new and exciting ideas with young people. He or she also needed to be willing to endure long hours of astronaut training. Above all, NASA felt their astronaut-teacher needed to understand that even though accidents were extremely rare, space travel could be dangerous.

While the Space Administration began its search for the person who would become America's first teacher in space, Christa McAuliffe was busy with her history students at Concord High School in Concord, Massachusetts. Among the classes she taught at Concord High was one she had created herself, called "The American Woman." Students in Christa's American woman course studied the role of women in the development of the United States. Women had always been essential contributors to American history, Christa taught her students, even when historians had not properly acknowledged their roles.

By the time Christa began teaching at Concord High in 1982, she had already devoted 12 years of her life to education. Her work with the students of Concord High was far from her only teaching activity, however. She taught classes in Christian doctrine at her church and served as a board member for several important teacher organizations. McAuliffe's leadership skills in education were widely admired.

In 1976, Christa gave birth to her first child, Scott Corrigan McAuliffe. Scott was joined by a baby sister, Caroline Corrigan McAuliffe, in 1979. While

The happy moment on July 19, 1985: hearing of her selection to be the first American teacher in space, McAuliffe hugs fellow teacher David Marquart.

Christa juggled her roles as mother and teacher, her husband, Steven, was building a highly successful career as an attorney.

Before fall 1984, joining the astronaut corps was unthinkable to this 36-year-old schoolteacher from Massachusetts. In November, however, Christa discovered NASA's plan to send the best-qualified teacher available into space. The idea of using the space shuttle as a classroom delighted her. She immediately decided to apply.

Christa McAuliffe was hardly alone in her decision to apply for the Teacher In Space Program, or TISP, as it was known within NASA. More than 10,000 of America's finest teachers jumped at the chance to teach from NASA's orbiting classroom. NASA's expert panels studied each application carefully. After months of research, they had whittled down the 10,000 applications to just 114, then to 10. Christa McAuliffe passed each cut with flying colors. On July 19, 1985, the world discovered that the United States had a brand new astronaut; Christa McAuliffe would be America's first teacher in space.

How did she feel about being selected over so many talented teachers? "I'm still kind of floating," she told reporters after the announcement, "I don't know when I'll come down to earth." Part of McAuliffe's success as an applicant came from her belief that America's space explorers were making the kind of fantastic discoveries that excited students about the universe around them. Inspired by female astronauts like Sally Ride, Christa McAuliffe also saw the space program as providing future opportunities for women and men alike.

Now that she had set her sights on the stars, McAuliffe had much to learn about the machine that would take her there. The space shuttle, as she quickly realized during her astronaut training, was

McAuliffe (left) trains for the Challenger *mission on a "zero gravity" aircraft. With her is backup Barbara Morgan.*

the most complex machine ever built. She spent long hours studying how astronauts flew the shuttle safely into space, and what to do in case of an emergency. She also needed to learn the many special ways of doing ordinary things in space. Eating a meal was not as simple in zero gravity as it was on Earth! Christa practiced living in space with fellow teacher Barbara Morgan, her stand-in. If Christa became ill and could not make her historic flight, Barbara Morgan would take her place.

Working closely with NASA, Christa put together a lesson plan to teach from her orbiting classroom. She would teach two classes from the space shuttle. Each would be broadcast to classrooms across the country. During the first lesson, mission specialist McAuliffe would take her students on a special tour of the space shuttle, explaining how astronauts lived and worked in space. During her second lesson, she would discuss the many benefits of space exploration and outline goals for future astronauts to achieve. She would also tape several lessons to be broadcast after her return to Earth.

Christa McAuliffe would have plenty of help preparing her lessons in orbit. For the trip, space shuttle *Challenger* would carry a full crew of seven astronauts. Besides America's first teacher in space, *Challenger* would carry mission commander Francis Scobee, pilot Michael Smith, and mission specialists Judith Resnik, Ronald McNair, Gregory Jarvis, and Ellison Onizuka.

Commander Scobee was happy to have Christa as a member of *Challenger*'s crew. He felt Christa's presence aboard the shuttle gave their mission a special importance. While Christa talked to her thousands of students below, *Challenger*'s crew would launch a new communications satellite and conduct a number of scientific experiments.

Christa's flight was known to NASA's engineers as Mission 51-L. Thousands of skilled technicians had spent months preparing *Challenger* and its cargo for Mission 51-L. After each shuttle returned from orbit, it was inspected inside and out. Every computer, every hatch, every knob, switch, and button aboard the shuttle had to work perfectly before the shuttle could be cleared for launch. As launch day approached, fresh food and clean clothing for the astronauts were loaded aboard. *Challenger* was ready to go.

The crew was, too. On the morning of the launch, January 28, 1986, McAuliffe and her crewmembers sat down for the traditional astronauts' breakfast of steak and eggs. Everyone, especially Christa, was excited about their mission. They had practiced repeatedly. Now they would show America what skilled astronauts could do.

But some NASA engineers worried about the weather. Although the skies were clear, the temperature had dropped below freezing during the night. The engineers worried that the cold could have damaged important parts of the shuttle's booster rockets. These rockets would help the shuttle lift off from the launching pad, and then they would drop into the ocean as the shuttle roared on toward space. NASA decided the cold weather posed no real danger. The launch would continue as planned.

Christa McAuliffe and the crew of the *Challenger* lifted off from Launch Pad 39B at Cape Canaveral, Florida, at exactly 11:38 in the morning. Several miles from the launch site, families of the crew, including Christa's husband, Steven, and their two children, watched as *Challenger* rose gracefully into the sky. But 73 seconds into the flight, something went terribly wrong. Space shuttle *Challenger* disappeared in a gigantic explosion. The shuttle, the satellite in the cargo bay, the food and clothes carefully packed away, and seven of America's finest astronauts were gone forever.

In the years after the disaster, Americans tried to make sense of what had happened. It was later shown that one of the shuttle's booster rockets had been damaged by the cold weather. The damaged rocket had caused the accident, just as some engineers had worried the day of the launch.

Although the country was terribly saddened by the loss of *Challenger* and its crew, most Americans felt that space exploration must continue. A new space shuttle was built to replace the *Challenger*, and new astronauts stepped forward to take the places of those who had been lost. Christa McAuliffe's sacrifice to the cause of education has not been forgotten. A planetarium and several schools have since been named after her. And the good work she started goes on and will go on among future generations of teachers, astronauts, dreamers, and explorers.

The tragedy that shocked the nation: less than two minutes after takeoff, Challenger *explodes.*

EILEEN M. COLLINS

Although he could not yet see the space shuttle itself, cosmonaut Alexander Viktorenko knew *Discovery* was close. Every few minutes now, Viktorenko could see the silent white flash of *Discovery*'s small thruster rockets igniting against the black background of space. The shuttle astronauts were using their thrusters to slow *Discovery* down as it approached the giant Russian space station *Mir*.

Soon the familiar shape of the space shuttle loomed large enough for cosmonauts aboard *Mir* to make out *Discovery*'s commander and pilot sitting behind the shuttle's large cockpit windows. In the lefthand cockpit seat sat *Discovery*'s commander, astronaut James D. Wetherbee. Wetherbee, an experienced Navy pilot, had already completed two space missions. Beside him sat another experienced military pilot, Eileen Collins, making her first space flight. As the space shuttle's first female pilot, Collins was also making history.

Eileen Marie Collins was born on November 19, 1956, in Elmira, New York, the daughter of James and Rose Marie Collins. As a young girl, she was fascinated by flying

Eileen Collins finishes a training exercise in May 1997, the day before her second space flight.

47

machines. Collins attended a summer camp near a landing field for airplanes without engines, called gliders. With their long, slender wings, gliders struck her as graceful and beautiful in the air. She often watched them wheeling high overhead in the summer sunshine. Young Eileen Collins was soon captivated by the freedom and excitement of flight.

Although she longed to become a pilot, her family could not afford the expensive lessons. With early adulthood came a job, and Collins found herself with some money of her own. She began paying for her own flying lessons at age 19.

After graduating from Elmira Free Academy in 1974, Collins began her long college career at Corning Community College. She was awarded an associate's degree in mathematics and science in 1976. Two years later, she earned a bachelor's degree in mathematics and economics at Syracuse University in Syracuse, New York. Still later, she would finish not one, but two, master's degrees.

While at Syracuse University, Eileen Collins enrolled in the air force's Reserve Officers' Training Corps, or ROTC. She hoped the air force would someday train her to become a professional pilot. First, there were many difficult courses to pass. In addition to her regular classes, Collins attended special air force classes. She learned how the air force protected the country during war, and about the many types of aircraft she might someday fly as an air force pilot. On certain days of the week, Collins and her ROTC classmates wore their air force uniforms to class.

When her work at Syracuse University was complete, Collins was awarded an officer's commission in the air force. She was now Second Lieutenant Eileen Collins. The air force decided to develop Eileen's piloting skills. She was assigned to the air force's Undergraduate Pilot Training Program at Vance Air Force Base in Oklahoma. She spent many long days and nights improving her flying abilities. Eileen Collins did not want to be second best at anything she tried. By 1979, she had completed the rigorous course and had earned her wings as an air force pilot.

Next, she became an instructor pilot herself at the flight training school, teaching new air force pilots to fly the T-38 jet trainer. Her assignment as a flying instructor showed how far her flying skills had developed. Now Eileen Collins was teaching others the delicate art of flying.

After several years of T-38 instruction in Oklahoma, Eileen was ready for a bigger challenge. The Lockheed C-141 Starlifter was one of the world's largest transport aircraft. With its four engines and enormous interior, the Starlifter could carry hundreds of soldiers or tons of supplies to any destination in the world. Between 1983 and 1985, Collins served as both a Starlifter commander and instructor. When she was not carrying out important missions around the globe, she was busy teaching new Starlifter pilots how to safely handle the tremendous cargo plane.

Collins's teaching skills were widely admired throughout the air force. After finishing her assignment as a Starlifter pilot, she taught mathematics to air force cadets at the Air Force Academy in Colorado. The air force continued to find new challenges for Eileen. With her teaching assignment completed, she was sent back to school as a student once again, this time at the air force's elite test-pilot training facility in California. Test pilots were given the dangerous duty of flying new kinds of airplanes for the first time. The test-pilot school was even located in the middle of a desert so that no one on the ground would be hurt when an airplane crashed.

The space shuttle Discovery *lifts off on February 3, 1995– Collins's first space flight.*

In January 1990, while Collins was finishing her test-pilot training, she received the exciting news that NASA had selected her to become an astronaut. Although other women had served as astronauts and as shuttle crew members, NASA needed Collins for a special role aboard the shuttle. Would she consider becoming America's first female space shuttle pilot? Eileen Collins didn't hesitate to say yes.

She relocated to Houston, Texas, to begin her long training as a space shuttle pilot. Flying the shuttle was very different from flying conventional airplanes. Although the shuttle wasn't as big as her Starlifter had been, it was much more complicated. Despite the many differences, some aspects of flying the space shuttle were familiar. She would be delivering important cargoes to far-off locations, just like she had done so many times in the C-141. Of course, space was a bit farther off than anyplace she'd flown before! Collins also learned that the space shuttle glided back through Earth's atmosphere without the help of an engine after its mission in space was done. After spending so many days watching gliders at summer camp, Eileen Collins would finally be getting one of her own.

On February 3, 1995, NASA's launch pad technicians helped Collins into her bulky flight suit. Out on Launch Pad 39-B, space shuttle *Discovery* waited in the cool night air for its 20th trip into space. At exactly 22 minutes after midnight, *Discovery*'s main engines roared to life. Eileen Collins's first journey into outer space had begun.

Shuttle mission STS-63 called for Collins and her crewmates to accomplish several important goals. The primary objective involved practicing a difficult operation known as docking. Future shuttle astronauts would gently connect their spacecraft to the

Russian space station *Mir,* allowing astronauts and supplies to be carried back and forth between the two. At NASA's astronaut training facility in Houston, Texas, Collins and Shuttle Commander Wetherbee had practiced docking many times. While cosmonaut Viktorenko watched from *Mir,* Collins and Wetherbee carefully guided *Discovery* to within just 37 feet of the spidery Russian space station. They had proven that future shuttle astronauts could safely dock with *Mir.*

Eileen Collins's experience with the intricate docking maneuver would be put to the test on her second space flight. Mission STS-84 lifted off from Cape Canaveral on May 15, 1997. Its primary objectives: deliver astronaut Michael Foale to *Mir* and bring *Mir* crewmember Jerry Linenger back home to Earth. Collins again flew as pilot, with veteran astronaut

At the pilot's station in Discovery, *Collins prepares for a "hotfiring" procedure shortly before the space shuttle's rendezvous with* Mir *in February 1995.*

Between flights, Collins enters Atlantis *for a dry run in late 1997. The next March, she was named the first U.S. woman space shuttle commander.*

Charles Precourt serving as mission commander. With great skill, Collins and Precourt safely docked space shuttle *Atlantis* with *Mir,* delivering their passenger and nearly 4,000 pounds of supplies. NASA judged their 10-day flight a complete success.

Like the air force, NASA was determined to keep challenging Eileen Collins. On March 5, 1998, at a special White House ceremony, first lady Hillary

Rodham Clinton made a historic announcement. Eileen Collins was to be America's next space shuttle commander, the first time ever a U.S. space mission would be commanded by a woman. Eileen Collins had occupied the space shuttle's righthand cockpit seat on two space missions. But for Mission STS-93, she would claim the lefthand seat—the commander's seat. "When I was a child," Collins said, "I dreamed about space. . . . It's my hope that all children, boys and girls, will see this mission and be inspired to reach for their dreams, because dreams do come true."

Eileen Collins, now a lieutenant colonel in the U.S. Air Force, and the mother of a young daughter, would realize her dream in December 1998, when space shuttle *Columbia* was scheduled to roar into orbit.

MAE C. JEMISON

Many people around the world know Dr. Mae C. Jemison as the first African-American woman to venture into space. But there is much more to understanding who Mae Jemison is. She is a doctor, a college professor, a business woman, a humanitarian, and a role model. Perhaps most important, Mae Jemison is a genuine American hero.

At age 16, when most of her friends were learning to drive a car, this future astronaut entered college. By her 21st birthday, Jemison had earned not one but two university degrees. Just four years later, she had earned a medical degree and the title "doctor." In June 1987, the National Aeronautics and Space Administration (NASA) selected Dr. Mae Jemison to become a space-shuttle astronaut. In 1992, Dr. Jemison climbed aboard the space shuttle *Endeavor* and rocketed into orbit, the first African-American woman in space.

Mae C. Jemison was born on October 17, 1956, in Decatur, Alabama. She was the youngest of Charlie and Dorothy Jemison's three children. When Mae was still quite young, the

Mae Jemison examines the Spacelab in the space shuttle Endeavor *before her historic flight in 1992.*

family moved to Chicago. Her father worked as a maintenance supervisor, and her mother taught English and mathematics at an elementary school.

Mr. and Mrs. Jemison encouraged Mae, her brother, Charles, and her sister, Ada, to set their sights high in life and to work hard. As a young girl, Mae spent many hours in the library reading, studying, and thinking. Surrounded by books full of exciting ideas, she thought about the ways modern inventions could be used to make the world a better place. She read of how doctors had developed medicines to cure diseases, and how inventors and engineers had created airplanes and telephones to connect people in distant countries. Mae began to see that science could both harm and benefit humanity. She liked the idea of using science and modern technology to help people in faraway places solve their problems.

As a black woman, Jemison faced special challenges as she worked to become a doctor and an astronaut. Almost all astronauts, and even most doctors, were white men. Breaking into their world was not easy for Mae. Some people she encountered along the way felt that women—especially black women—had no place in the worlds of science and medicine. Jemison has spent much of her life proving they were wrong.

Even when Mae was still very young, people discouraged her dreams. In Chicago, Mae told her kindergarten teacher that she hoped to become a scientist someday. Her teacher told the little girl that she should become a nurse, a choice the teacher thought more suitable for girls. But Mae did not give up. "I was stubborn," she said later. In time, that stubbornness proved to be one of the qualities that brought her success.

In 1973, Mae Jemison graduated from Morgan

Park High School in Chicago, where her high grades won her a full scholarship to Stanford University in California, one of the nation's top universities.

Mae's academic success continued at Stanford. While most of her classmates completed just one college degree during their four years at the university, Mae Jemison earned two degrees in two very different fields: chemical engineering and African-American studies. She picked chemical engineering because she would need a strong background in chemistry to become a doctor. She picked African-American studies in order to find out more about her own heritage as an African American. She would later combine these interests in a way that excited her and gave her great satisfaction.

Mae took pride in her family's black heritage. She drew strength from her family history and from the many stories of black achievement in the United States. She resolved never to let racism stop her from realizing her dreams.

After graduating from Stanford, Mae enrolled as a medical student at Cornell University, which has one of the country's best medical schools. There Mae's professors encouraged her to study harder than she ever had before. After four years of intense work, Mae was awarded her medical degree in 1981. She was now Dr. Mae Jemison.

Mae Jemison had always wanted to use modern technology to help people throughout the world live better lives. Her interest in her ancestral homeland led her to West Africa, where she worked with the Peace Corps as a doctor. She found this to be a perfect way to combine her medical skill with her fascination with her African heritage. (The Peace Corps had been founded by President John F. Kennedy during Mae's early childhood. Peace Corps volunteers worked in many of the world's poorest coun-

On board the Endeavor, *Jemison wears a headband and other scientific monitoring gear.*

tries, helping people improve their lives.) Mae Jemison doctored other Peace Corps volunteers, helping them stay healthy so they could do their jobs.

In 1985, back in the United States, Dr. Jemison continued her medical work in Los Angeles. But she soon came up with a new challenge, as she began thinking about the power of modern technology to improve life on Earth. With its tremendous scientific and technological resources, she thought, the Space Administration would be the perfect partner in her fight to use science for peaceful purposes. A year after she returned from Africa, Mae Jemison made an exciting decision: she would send off an application to NASA's astronaut program.

For its 1987 astronaut class, the space administration received nearly 2,000 applications. Many of the applicants, including Dr. Jemison, represented the best minds in their fields. When NASA announced the new astronauts, Mae's name was on the list. She was elated.

Aboard the space shuttle, Dr. Jemison would serve as a space scientist. She would be responsible for carrying out important scientific experiments while the shuttle orbited Earth. Scientists expected many of the experiments to answer a number of important life science questions. As a doctor, Mae felt a special excitement about studying how the human body functioned in space. She wondered if her experiments might improve the medicines and medical techniques used every day on Earth.

Before NASA would include her on the space-shuttle team, Dr. Jemison had to complete the demanding astronaut training program. Once again, she found herself studying hard. She memorized the many emergency drills—what to do in case any one of a number of things went wrong with the space shuttle. She learned how to eat, drink, and dress under zero-gravity conditions, when her glass and fork and even her food could float in the air, and her stomach couldn't tell down from up. Of course, she spent endless hours learning how to perform her scientific experiments. Many scientists and doctors on Earth would depend on Mae to do her experiments correctly. Otherwise, any scientific information Mae gathered in space would be useless.

On September 12, 1992, the space shuttle *Endeavor* blasted off from Pad 39–B at NASA's Kennedy Space Center in Florida. Aboard were seven crew members, including Mamoru Mohri, the first Japanese astronaut to ride into space on a U.S. space shuttle. Like many people who make history, Dr. Mae Jemison was almost too busy with her work to reflect on her achievement as the first female African American in space.

Dr. Jemison performed most of her scientific experiments inside the Spacelab, a giant laboratory that remained inside the shuttle's cargo bay during the entire flight. With one of her most important experiments, scientists hoped to discover a new way to prevent future astronauts from getting space sick, a common aerospace ailment. Space sickness, a kind of nausea, affected about half of NASA's astronauts during their first days aboard the shuttle. Dr. Jemison knew that space sickness probably resulted from the lack of gravity in space. Without gravity, the body cannot tell which direction is up, leaving many astronauts with upset stomachs.

NASA asked Dr. Jemison to try out a new and unusual cure for space sickness that did not involve the use of medicine. Mae was instructed to use the power of her mind to calm her body so that she would not feel ill in space. Dr. Jemison hoped that this kind of cure for sickness—a cure without medicine—might be useful on Earth as well as in space.

In a Spacelab experiment, Jemison injects fluid into a mannequin's hand.

Another experiment looked at how animals adjust to life in space. Aboard the Spacelab were eggs taken from South African frogs. NASA's scientists wanted to know if the eggs would grow normally in zero gravity. Dr. Jemison found that they would. "It was exciting," she later said, "because that's a question we didn't have any information on before."

While in space, Dr. Jemison also observed the behavior of other kinds of animals in zero gravity, including fruit flies and Japanese fish called *koi*.

Eight days after blasting off from sunny Florida, the *Endeavor* brought Dr. Jemison and her crewmates safely back to Earth. During its many orbits around the earth, the shuttle had traveled more than three million miles. Dr. Jemison and her fellow astronauts had carried out more than 40 important science experiments. The information they collected would help scientists develop new technologies, medicines, and materials for people on Earth.

After her space flight, Mae Jemison decided to leave the space agency. Her decision to resign from

NASA after only one flight into space probably didn't surprise the people closest to Mae. They know that once she has met a challenge—such as becoming an astronaut—she's usually ready to move on to the next goal in her life.

Since her history-making voyage into space, Dr. Jemison has continued to concentrate on one of her most important goals: using science to help people. To work toward this goal, Dr. Jemison relies on a research corporation she created called the Jemison Group. Scientists at the Jemison Group share Dr. Jemison's special vision and her compassion for people in need around the world.

In the future, the Jemison Group hopes to construct a satellite system called the Alafiya. *Alafiya* is an African word meaning "good health." By using satellites in space, Alafiya will allow people in remote parts of Africa to communicate with doctors in other parts of the world. West Africans will be able to speak to their doctors from thousands of miles away.

Is Mae Jemison angry at the people who tried to discourage her from realizing her dreams because she is a black woman? Not at all. She's been happy to prove their theories wrong simply by living her life. It's just the kind of challenge any good scientist loves. Had she accomplished just one of her many goals—becoming either a doctor, or an astronaut, or a college professor—Dr. Mae Jemison's professional life could be called a success. But she has done it all. She has also done it against the longest odds. She has used her life to prove that one person with courage has the power to make the world a better place.

CHRONOLOGY

1784   Frenchwoman Elisabeth Thible becomes the first woman to fly, ascending via balloon.

1886   American balloonist Mary Myers sets world altitude record of four miles.

1910   Baroness Raymonde de Laroche becomes the first woman to be awarded a pilot's license; she is killed in a crash nine years later.

1911   Famed U.S. aviator Harriet Quimby shows her flying skills—and her purple satin flying suit—to an enthralled public.

1912   Harriet Quimby becomes the first woman to cross the English Channel by air and is killed in a crash that same year.

1916   Ruth Law sets a nonstop cross-county speed record by flying from Chicago to Hornell, New York; later she unsuccessfully applies for aerial combat duty with the U.S. army.

1921   Celebrated African-American aviator Bessie Coleman earns her pilot's license; she dies five years later while rehearsing for an air show.

1928   Amelia Earhart flies the Atlantic as a passenger.

1930   Ruth Nichols sets a new women's coast-to-coast speed record in a borrowed Lockheed Vega.

1932   Amelia Earhart becomes the first woman to fly the Atlantic solo.

1937   Amelia Earhart and Fred Noonan attempt round-the-world flight but disappear over the Pacific Ocean.

1943   The Women Airforce Service Pilots (WASP) program is created; WASP pilots ferry U.S. aircraft from factories to domestic combat dispersal points throughout World War II.

1963   Russian cosmonaut Valentina Tereshkova becomes the first woman in space.

1968   Elinor Williams is named the nation's first African-American air traffic controller.

1983   America's first female astronaut, Sally Ride, rockets into orbit.

1988   Barbara Barrett is named deputy administrator of the Federal Aviation Administration.

1992   America's first female African-American astronaut, Mae Jemison, orbits the Earth.

1998   Eileen Collins is named America's first female space shuttle commander.

FURTHER READING

Asimov, Issac. *The Red Planet: Mars.* Milwaukee: Gareth Stevens, 1994.

Behrens, June. *Sally Ride: An American First.* Chicago: Children's Press, 1984.

Billings, Charlene. *Christa McAuliffe: Pioneer Space Teacher.* Hillside, N.J.: Enslow, 1986.

Boyne, Walter. *The Smithsonian Book of Flight for Young People.* New York: Aladdin, 1988.

Chadwick, Roxane. *Amelia Earhart: Aviation Pioneer.* Minneapolis: Learner, 1987.

Embury, Barbara. *The Dream Is Alive: A Flight of Discovery Aboard the Space Shuttle.* New York: Harper & Row, 1990.

Fox, Mary. *Women Astronauts: Aboard the Shuttle.* New York: Messner, 1984.

Nahum, Andrew. *Flying Machine.* New York: Knopf, 1990.

Ride, Sally, and Okie, Susan. *To Space and Back.* New York: Lothrop, Lee, & Shepard, 1986.

Shore, Nancy. *Amelia Earhart.* Philadelphia: Chelsea House, 1987.

Vogt, Gregory. *Mars.* Brookfield, Conn.: Millbrook, 1994.

——. *Spaceships.* New York: Franklin Watts, 1990.

Wood, Leigh Hope. *Amelia Earhart.* Philadelphia: Chelsea House, 1997.

Yannuzzi, Della. *Mae Jemison: A Space Biography.* Springfield, N.J.: Enslow Publishers, 1998.

INDEX

ABOUT THE AUTHOR

Doug Buchanan grew up in Charlotte, North Carolina. He holds a bachelor's and a master's degree in history from the University of Mississippi. He is currently director of a history museum in Albemarle, North Carolina. This is his first book for Chelsea House.